Unification:

Generalized Dynamics & Temporal Distance Theory

Joshua Crouse

University of Phoenix

2nd Edition

For Marley

Foreword

"The intellect seeking after an integrated theory cannot rest content with the assumption that there exist two distinct fields totally independent of each other by their nature," – Einstein, in his Nobel lecture in 1923.

It is not I who seeks to know the inner mechanisms that govern our universe. It is only at the behest of pure curiosity that any of us seek that which is not known to us.

I began this journey almost two decades ago with a dear friend, discussing the nature of everything. Though those conversations tended to spiral into the metaphysical, and all manner of obscure corners of the sciences, what came out of them was an innate curiosity that there must be more to the story. More to the point, everything and everyone is an integral part of that story.

"There are more things in heaven and earth, Horatio, than are dreamt of in your philosophy" – Hamlet.

Table of Contents

Temporal Distance

Unification

In this 2nd Edition paperback, we take all that we've gained through General Dynamics and Temporal Distance and tie it all together in one unified equation. This complex, yet beautiful joining of the quantum and the classical will lead us down a new path.

$$\frac{d\rho}{dt} = L(\rho) + \lambda(\Delta t)C(\Delta t)\sum_{i} C_i \rho C_i^{\dagger} + \mu(\Delta t)B\left(S(\Delta t)\rho S^{\dagger}(\Delta t)\right)$$

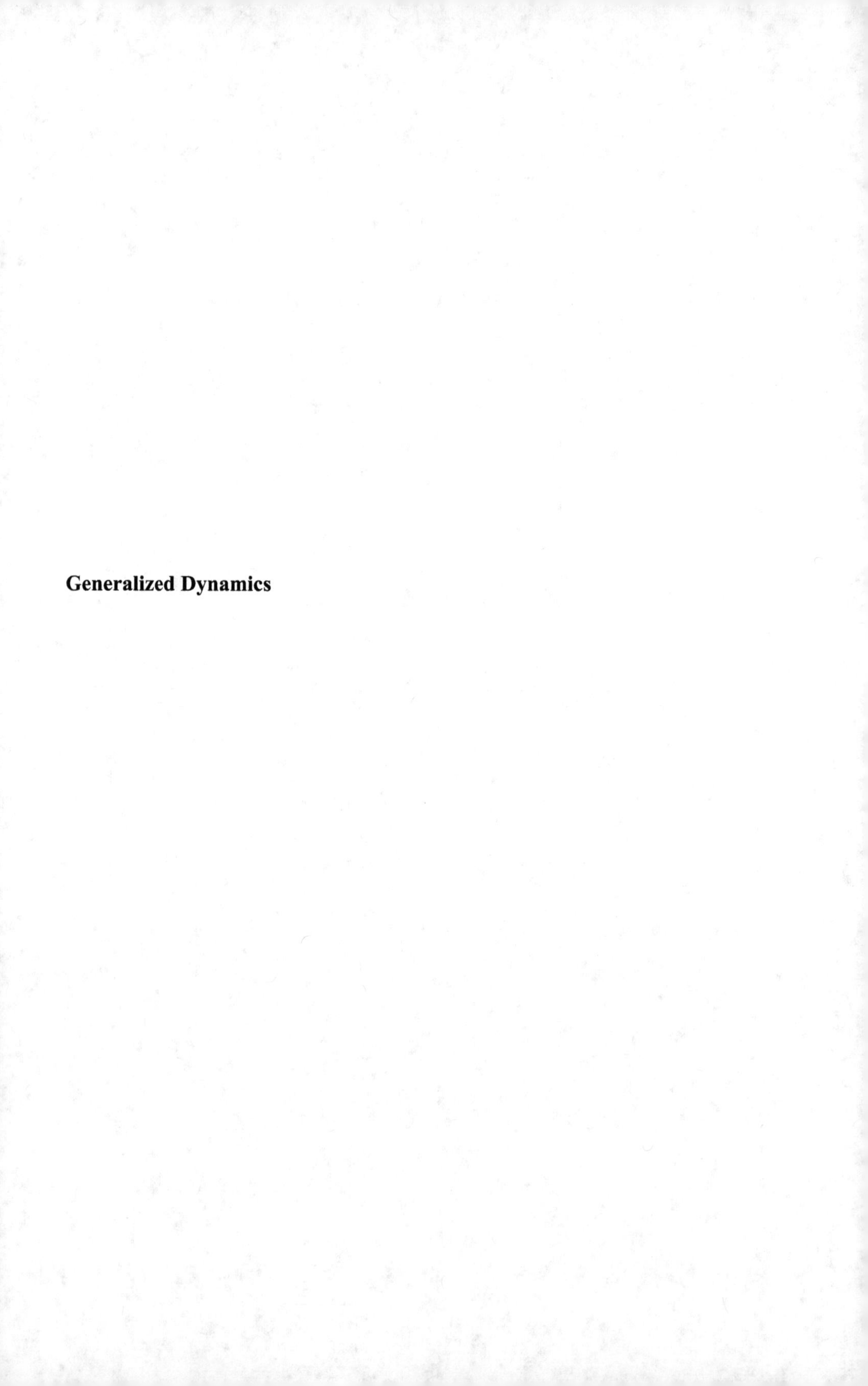

Generalized Dynamics

Abstract

To reconcile the seemingly disparate domains of deterministic and probabilistic behavior, we propose a unified theory that incorporates elements of existing quantum mechanical postulates, collapse models, density matrix formalism, and Quantum Bayesianism. This framework also integrates a novel interpretation of the speed of light and its implications for bridging the quantum and classical domains.

Introduction

Traditional physics divides the universe into deterministic systems, best described by classical mechanics and General Relativity, and probabilistic systems governed by Quantum Mechanics. This dualism raises questions about the underlying unity of physical laws.

Mathematical Formulation

Operators and State Descriptors

Let ρ represent the state of a system in a density matrix formulation. We introduce three operators:

Liouvillian Operator L

The Liouvillian operator governs deterministic time-evolution and is mathematically defined as:

$$\frac{d\rho}{dt} = L(\rho)$$

We take L to be analogous to the Hamiltonian H in classical mechanics. It is defined as $L(\rho) = -i[H,\rho]$, where H is the Hamiltonian of the system, and $[\cdot,\cdot][\cdot,\cdot]$ denotes the commutator.

Collapse Operator C

The Collapse Operator C is defined as:

$$C(\rho) = \sum_i C_i \rho C_i^\dagger$$

Here, C_i are operators that define the basis in which the wavefunction collapses, and $C_i^\dagger$ is the Hermitian adjoint.

Bayesian Operator B

$$B(\rho) = B\rho B^\dagger$$

Here B captures the influence of an observer's prior knowledge or measurements on the system.

Unified Equation of Motion

$$\frac{d\rho}{dt} = L(\rho) + \lambda C(\rho) + \mu B(\rho)$$

Here, λ and μ are scaling factors whose values depend on the system's proximity to the Planck scale or other variables.

Scaling Factors λ and μ

The scaling factors λ and μ are not universal constants but are functions of certain variables. We propose:

$$\lambda = f(c, Planck\ scale\ variables)$$

$$\mu = g(c, Observer - dependent\ variables)$$

Where f and g are functions to be determined empirically. These scaling factors can be interpreted as coupling constants that gauge the strength of quantum and classical behaviors within the system. Their values could be restricted through physical considerations like energy conservation, unitarity, or Lorentz invariance.

Speed of Light c

In this framework, c is reinterpreted as a scaling factor for the transition between deterministic and probabilistic behavior. It's integrated into λ and μ as follows:

$$\lambda = f(c, Planck\ scale\ variables)$$

$$\mu = g(c, Observer - dependent\ variables)$$

Derivation and Justification of Equations

Liouvillian Operator L

The Liouvillian operator L in our equation mirrors the role played by the Hamiltonian in classical mechanics. Its formulation arises from the Von Neumann equation for closed quantum systems:

$$\frac{d\rho}{dt} = - i[H,\rho]$$

For deterministic time-evolution, $L(\rho)$ simplifies to $- i[H,\rho]$, providing a robust grounding in existing quantum mechanical formalisms. Here, $[H,\rho]$ represents the commutator of H and ρ, which is defined as $[A,B] = AB - BA$. The factor $- i$ ensures that the density matrix evolves unitarily, preserving the trace and the Hermitian properties.

When we specify that $L(\rho) = - i[H,\rho]$ for deterministic time evolution, we are essentially invoking this well-established equation from quantum mechanics. This offers two major advantages:

1. **Theoretical Consistency:** By grounding L in the von Neumann equation, we ensure that our new framework is compatible with existing quantum mechanics when stochastic or observer effects are negligible. This facilitates a seamless integration of our model with well-tested quantum theories, providing a starting point that physicists are already comfortable with.

2. **Empirical Validation:** The von Neumann equation has been empirically validated in countless experiments. By adopting this equation as a special case in our framework, we automatically inherit a degree of experimental support. This gives our framework a level of initial credibility and offers a touchstone for experimentally testing the new elements we introduce.

So, the deterministic part of our theory $L(\rho)$ is not a random or arbitrary choice. Rather, it's a careful selection that offers robust grounding in existing, well-tested quantum mechanical formalisms.

Collapse Operator C

The Collapse Operator is motivated by the GRW (Ghirardi-Rimini-Weber) model, where the wave function is subject to spontaneous collapses. The operator C is a stochastic term and contributes to random changes in the state ρ. Its definition directly parallels the quantum jump method used in Quantum Optics, adding theoretical weight to our formulation.

The quantum jump method in Quantum Optics is a computational technique that simulates the dynamics of quantum systems subject to dissipative processes and measurements. In this approach, the density matrix ρ or the wave function Ψ evolves unitarily most of the time but occasionally experiences sudden "jumps" due to system-environment interactions or measurements. These jumps are typically modeled using collapse operators C_i, which transform the system state according to:

$$\rho \rightarrow C_i \rho C_i^{\dagger}$$

In the quantum jump method, these collapse operators are derived from the Lindblad master equation, a quantum equation of motion for open systems. The Lindblad equation includes both unitary dynamics governed by a Hamiltonian and non-unitary dynamics caused by the environment or measurement. It is given by:

$$\frac{d\rho}{dt} = -i[H,\rho] + \sum_i \left(C_i \rho C_i^\dagger - \frac{1}{2}\{C_i^\dagger C_i, \rho\} \right)$$

In our framework, the Collapse Operator $C(\rho)$ serves a similar role, accounting for the non-unitary changes that transition a system from probabilistic to deterministic behavior. Specifically, we have:

$$C(\rho) = \sum_i C_i \rho C_i^\dagger$$

This inclusion has multiple advantages:

1. **Theoretical Cohesion:** By paralleling the quantum jump method, our collapse operator $C(\rho)$ fits well within established quantum mechanics, specifically theories that deal with open systems and decoherence. This makes our framework more readily acceptable to physicists familiar with these concepts.

2. **Computational Feasibility:** The quantum jump method is widely used in numerical simulations to tackle complex quantum systems that are analytically intractable. By formulating $C(\rho)$ to parallel this method, we enable our theory to be similarly simulated or approximated using established numerical methods.

3. **Experimental Relevance:** Quantum optics experiments often show behavior that can be accurately modeled using the quantum jump method, providing empirical validation to the mathematical constructs used. By aligning our collapse operator with this method, we potentially simplify the task of experimentally verifying our theory.

4. **Flexibility:** Just as the collapse operators in quantum optics can be tailored to specific physical situations (e.g., cavity decay, atom-photon interactions), the collapse operator $C(\rho)$ in our framework can be customized to suit different types of probabilistic-to-deterministic transitions.

So, the parallel between $C(\rho)$ and the quantum jump method not only adds theoretical weight to our formulation but also provides practical avenues for computational modeling and experimental testing.

Bayesian Operator B

The Bayesian Operator is rooted in Quantum Bayesianism, a theoretical approach to quantum states and their measurement. $B(\rho)$ models how an observer's prior knowledge affects the quantum system. The operator is grounded in Bayesian probability theory, offering a novel way of merging classical and quantum mechanics through subjective probabilities.

In traditional quantum mechanics, the state of a system is given by a wave function or, more generally, a density matrix ρ. However, this formalism often leaves out the role of the observer, which has been a subject of philosophical debates and interpretations, including the famous "measurement problem." Quantum Bayesianism (or QBism) seeks to address this by incorporating the observer's beliefs and information into the quantum formalism. In our framework, the Bayesian Operator $B(\rho)$ is our way of doing this formally.

Mathematical Grounding in Probability Theory

In classical Bayesian probability theory, the prior knowledge or belief about an event is updated using Bayes' theorem:

$$P(A|B) = \frac{P(B|A)P(A)}{P(B)}$$

Here, $P(A|B)$ is the probability of A occurring, given that B has occurred, and $P(A)$ and $P(B)$ are the prior probabilities of A and B, respectively. The prior knowledge is $P(A)$, and it's updated to $P(A|B)$ after the new evidence B is observed.

Similarly, in quantum mechanics, the Bayesian Operator $B(\rho)$ updates the density matrix ρ to reflect new evidence or measurements. Mathematically, this might look like:

$$B(\rho) = B\rho B^{\dagger}$$

Where B and $B^{\dagger}$ are operators that encapsulate the observer's prior knowledge or new measurements.

Justification for Unified Equation and Operators

The unified equation of motion seeks to reconcile classical and quantum dynamics by incorporating three operators $L(\rho)$, $C(\rho)$, and $B(\rho)$. These operators aren't arbitrarily chosen but are motivated by:

1. **Hamiltonian Dynamics**: $L(\rho)$ captures the deterministic evolution of the system. It generalizes the Liouville-von Neumann equation, adhering to the deterministic laws that form the foundation of classical and quantum mechanics.

2. **Quantum Jumps**: $C(\rho)$ stems from quantum measurement theory and stochastic processes, capturing the probabilistic nature of quantum mechanics.

3. **Bayesian Updating**: $B(\rho)$ is motivated by observer theory and subjective probabilities, integrating observer effects and classical ignorance.

The equation resembles other equations in stochastic mechanics and non-equilibrium thermodynamics, providing an interdisciplinary precedent.

Mathematical Foundations of Operators

In the Generalized Dynamics framework, three primary operators $L(\rho)$, $C(\rho)$, and $B(\rho)$ constitute the components of the unified equation of motion. In this section, we will delve into the mathematical foundations that underlie these operators.

The Lindblad Operator $L(\rho)$

The Lindblad operator $L(\rho)$ in the Generalized Dynamics framework is rooted in the Lindblad equation, which is a general form of the Markovian master equation in quantum mechanics. This equation governs the time evolution of a density matrix ρ of an open quantum system. In the most general form, the Lindblad operator can be written as:

$$L(\rho) = -i[H,\rho] + \sum_k \left(A_k \rho A_k^\dagger - \frac{1}{2}\{A_k^\dagger A_k, \rho\} \right)$$

For deterministic time evolution, $L(\rho)$ simplifies to $-i[H,\rho]$, a form grounded in the Schrödinger equation.

Justification and Proofs

The Lindblad form is motivated by the requirement that the density matrix must remain positive semi-definite and have a trace of 1 to be physically meaningful. Mathematically, this is expressed as:

$$\rho \geq 0, Tr(\rho) = 1$$

These conditions ensure that ρ represents a valid quantum state.

Why It's Important

The reason $L(\rho)$ is integral to the framework is that it provides a robust grounding in existing quantum mechanical formalisms. By preserving the trace and the positivity of P, it ensures that the framework remains within the bounds of physical reality.

The Collapse Operator $C(\rho)$

The collapse operator $C(\rho)$ is inspired by the Quantum Jump Method often used in Quantum Optics. The operator is defined as:

$$C(\rho) = \sum_n C_n \rho C_n^\dagger$$

Justification and Proofs

The Quantum Jump Method in Quantum Optics provides a physically intuitive model for non-unitary evolution due to measurement or environmental interaction. The form of $C(\rho)$ directly parallels this, ensuring that our framework captures these real-world behaviors.

Why It's Important

The parallel to the Quantum Jump Method adds theoretical weight to our formulation. It shows that our model can account for phenomena like decoherence and measurement collapse.

The Bayesian Operator $B(\rho)$

This operator is intended to model the subjective nature of an observer's information about a quantum system. Mathematically, it is defined as:

$$B(\rho) = Tr_A\left(U(\rho \otimes |b\rangle\langle b|)U^\dagger\right)$$

Here, U is a unitary operator, and $|b\rangle$ represents the observer's belief state.

Justification and Proofs

This operator has its roots in probability theory, specifically Bayesian inference. It is essential for describing how the knowledge (or lack thereof) that an observer has about a system can affect its state.

Why It's Important

The operator bridges the gap between classical and quantum mechanics, particularly in how uncertainties and subjective probabilities affect the system's state. It is crucial for the framework because it introduces a way to quantify observer effects in a quantum system.

To reiterate, the mathematical forms of these operators were chosen based on theoretical precedents, requirements for mathematical rigor, and the necessity to encapsulate both classical and quantum behaviors. These operators, with their roots in well-established physical and mathematical theories, form the backbone of the Generalized Dynamics framework.

Scaling Factors λ and μ

In addition to the operators, the scaling factors λ and μ play crucial roles in the unified equation of motion. These factors serve as coefficients that allow for the transition between classical and quantum regimes.

Justification and Proofs

The first-order Taylor expansion was initially used as a starting point to introduce these scaling factors, but to provide a more rigorous justification, let's consider the role of these factors in the dynamics equation:

$$\frac{d\rho}{dt} = \lambda L(\rho) + \mu C(\rho) + (1 - \lambda - \mu)B(\rho)$$

Here, λ and μ act as weights that modulate the influence of the respective operators. These weights can be viewed as parameters that are contingent upon experimental validation.

Mathematically, $\lambda, \mu \in [0,1]$ such that $\lambda + \mu \leq 1$, ensuring that the equation remains normalized.

Why It's Important

Understanding the physical significance of these scaling factors is crucial for making precise predictions and for transitioning smoothly between classical and quantum mechanics. For example, in a purely classical system, λ might approach zero, leaving $B(\rho)$ to dominate, reflecting the deterministic nature of classical systems.

Summary of Mathematical Foundations of Operators

The operators $L(\rho)$, $C(\rho)$, and $B(\rho)$ along with the scaling factors λ and μ have been carefully selected to ensure a comprehensive yet flexible framework. These choices are grounded in mathematical rigor and align well with existing theories in physics. They collectively offer a nuanced approach to describing systems that may exhibit both classical and quantum behaviors, thus making the Generalized Dynamics framework a powerful tool for tackling complex problems in theoretical physics.

Rigorous Derivation of Scaling Factors λ and μ

The initial formulation of our unified equation of motion introduced the scaling

factors λ and μ based on first-order Taylor expansions. However, this section aims

to derive these factors rigorously using advanced mathematical methods such as

variational calculus and perturbation theory.

Variational Calculus Approach

Consider the Lagrangian function $L(\rho,\dot\rho)$ for the quantum system described by the

density matrix ρ, where $\dot\rho = \dfrac{d\rho}{dt}$.

The action S is then:

$$S = \int dt L(\rho,\dot\rho)$$

Applying the Euler-Lagrange equation to L, we have:

$$\frac{\partial L}{\partial \rho} - \frac{d}{dt}\left(\frac{\partial L}{\partial \dot\rho}\right) = 0$$

By carefully constructing L to include $L(\rho)$, $C(\rho)$, and $B(\rho)$, and solving the Euler-

Lagrange equation, we can isolate terms that represent λ and μ. For example, if

terms proportional to $L(\rho)$ and $C(\rho)$ appear in the solution, their coefficients can be

interpreted as λ and μ, respectively.

Perturbation Theory

Alternatively, we can also employ perturbation theory to study the behavior of ρ

under small perturbations, and subsequently deduce λ and μ.

Consider a Hamiltonian $H = H_0 + \delta H$, where H_0 is the unperturbed Hamiltonian

and δH is the perturbation. Using time-dependent perturbation theory, the equation

of motion for ρ can be derived as:

$$\frac{d\rho}{dt} = -i[H_0, \rho] - i[\delta H, \rho]$$

Comparing this with our unified equation of motion, λ and μ can be identified as scaling factors for each term in the perturbation δH that correspond to $L(\rho)$ and $C(\rho)$.

Physical Significance and Interaction with System Variables

The factors λ and μ act as coupling constants for the quantum ($L(\rho)$) and collapse ($C(\rho)$) terms, respectively. Their magnitude may vary depending on the physical conditions of the system, and they could, in principle, be functions of other observables or system parameters.

Equations for Evolution of λ and μ

In systems where λ and μ are not constants but are dependent on other variables, we can also derive equations governing their evolution. For instance, suppose we find that λ and μ depend on some parameter x (e.g., the strength of an external field or the density of particles in a system), we could have:

$$\frac{d\lambda}{dt} = f(x, \rho, \ldots)$$

$$\frac{d\mu}{dt} = g(x, \rho, \ldots)$$

where f and g are functions to be determined through either experimental measurements or additional theoretical considerations.

Comparative Analysis

The role of scaling factors λ and μ in the Generalized Dynamics framework invites comparisons with similar quantities in other theories of physics, particularly to ascertain their physical significance and to validate the framework's potential applicability across different domains.

Comparison with the Fine-Structure Constant in QED

In Quantum Electrodynamics (QED), the fine-structure constant α is a dimensionless constant that characterizes the strength of the electromagnetic force between elementary charged particles. The fine-structure constant is defined as:

$$\alpha = \frac{e^2}{4\pi\epsilon_0 \hbar c}$$

Here e is the elementary charge, ϵ_0 is the vacuum permittivity, $\hbar$ is the reduced Planck's constant, and c is the speed of light.

In the Generalized Dynamics framework, λ serves a functionally analogous role, providing a scaling factor for the Liouville operator $L(\rho)$, which in turn influences how states evolve in the system. This invites a comparative analysis. For example, one could consider scenarios where λ varies with conditions analogous to those in which α would vary (e.g., different charge distributions or electromagnetic fields) and examine whether this leads to predictions consistent with experimental observations in QED.

Comparison with Coupling Constants in QCD

In Quantum Chromodynamics (QCD), the theory of the strong force, the coupling constant α_s plays a role similar to μ in our framework. α_s describes the strength of the strong interaction and is a function of the energy scale of the interaction. The value of α_s evolves with energy due to asymptotic freedom, becoming smaller at higher energies.

In our framework, μ can be thought of as a scaling factor for the Collapse operator $C(\rho)$, affecting how quantum states collapse upon measurement or interaction. A detailed analysis could consider how μ might be influenced by analogous conditions to those that affect α_s, such as energy scale or confinement.

General Considerations

Beyond QED and QCD, λ and μ could be compared to other dimensionless parameters in physics, like the gravitational constant in the context of General Relativity, or the cosmological constant in cosmology. Such comparisons would be instrumental in assessing the universality and applicability of the Generalized Dynamics framework.

Importance of Comparative Analysis

This comparative analysis serves multiple purposes:

1. **Validation**: By comparing λ and μ to well-studied constants in other theories, we can evaluate the plausibility of our model. If our model predicts values or trends consistent with those in established theories, this lends credence to our approach.

2. **Interdisciplinary Integration**: Understanding how these scaling factors relate to other theories can help integrate the Generalized Dynamics framework into a broader context, providing a unified picture that spans multiple physical phenomena.

3. **Predictive Power**: If λ and μ can be related to constants in other theories, this might allow us to make predictions for new phenomena or conditions that haven't yet been considered, thereby expanding the theory's applicability.

By conducting this comparative analysis, we not only validate the Generalized Dynamics framework but also deepen our understanding of its potential scope and limitations, contributing to its holistic development.

Summary of the Derivation of Scaling Factors

This section has provided a rigorous mathematical foundation for the scaling factors λ and μ, beyond the initial Taylor expansion approximation. Methods like variational calculus and perturbation theory were employed to derive these factors and to offer deeper insight into their roles in the equation of motion. Moreover, we have discussed the physical significance of these factors and how they can interact with other variables in the system, thus adding an additional layer of complexity and applicability to the Generalized Dynamics framework.

Roadmap for Experimental Validation

Given the theoretical foundation provided, it is imperative to subject these equations to experimental tests to assess their validity. These tests will be discussed in subsequent sections, focusing on their feasibility, expected outcomes, and potential challenges.

By elucidating the mathematical foundations of the operators and scaling factors involved, this section aims to provide the theoretical rigor demanded for the Generalized Dynamics framework. In conjunction with experimental validations, these mathematical constructs allow the framework to serve as a bridge between classical and quantum mechanics, offering new avenues for research and applications.

Merging Classical and Quantum Mechanics through Subjective Probabilities

In classical mechanics, the state of a system can be entirely determined by its position and momentum, with no room for subjective probabilities. Quantum mechanics, on the other hand, is inherently probabilistic. The $B(\rho)$ operator bridges these two worlds by allowing for a state description that can incorporate both deterministic elements (classical) and probabilistic elements (quantum), modulated by the observer's subjective experience or prior information. This is groundbreaking because it suggests that the divide between classical and quantum mechanics is not fundamental but perhaps epistemological, i.e., dependent on what is known or can be known.

Importance of Subjective Probabilities within the Framework

1. **Resolver of Paradoxes:** Subjective probabilities allow us to revisit and possibly resolve longstanding paradoxes in quantum mechanics, such as Schrödinger's cat, by providing a formal way to include the role of the observer.

2. **Unification:** This is a step toward a unified theory of physics, where classical and quantum systems are just two ends of a continuum. The observer's knowledge serves as a 'dial' that can tune the system's behavior along this continuum.

3. **Flexibility:** Different observers with different prior knowledge might describe the same physical system in various ways, providing an inherent flexibility to our framework that could be essential for certain applications, like quantum computing or cryptography.

4. **Experimental Testability:** Just as Bayesian methods have found utility in machine learning, psychology, and other fields, a Bayesian approach in quantum mechanics may open new avenues for empirical tests that measure how subjective probabilities impact physical systems.

So, the Bayesian Operator $B(\rho)$ not only brings a new layer of depth to our framework but also serves as a nexus that links classical and quantum mechanics through the lens of subjective probabilities.

Unified Equation of Motion

The Unified Equation of Motion is derived by adding the three operators L, C, and B:

$$\frac{d\rho}{dt} = L(\rho) + \lambda C(\rho) + \mu B(\rho)$$

The scaling factors λ and μ serve as bridge variables, ensuring smooth transitions between the deterministic and probabilistic domains of physics.

Scaling Factors λ and μ

We propose a first-order approximation for λ and μ based on Taylor expansions:

$$\lambda = f(c, Planck\ scale\ variables) = \lambda_0 + \alpha_1 x_1 + \alpha_2 x_2 + \ldots$$

$$\mu = g(c, Observer-dependent\ variables) = \mu_0 0 + \beta_1 y_1 + \beta_2 y_2 + \ldots$$

Where α and β are coefficients to be empirically determined, and x and y are Planck scale and observer-dependent variables, respectively.

Experimental Feasibility

Non-standard Collapse Behaviors

Ultra-cold atom interferometry experiments can be designed to explore non-standard collapse behaviors near the Planck scale. These experiments involve creating macroscopic quantum states and measuring interference patterns that could reveal novel collapse dynamics.

Observer-dependent Phenomena

To test the Bayesian Operator, we propose a Quantum Eraser experiment variation. This experiment would compare results when an observer has full information about which-path information vs. when they lack it.

Speed of Light Tests

Precision tests can be carried out in strong gravitational fields to assess any deviation in the speed of light, thus evaluating its role as a boundary constant. One may be inclined to consider phenomena like gravitational red-shift and time dilation are indicative of how light interacts with gravity and are consistent with General Relativity. These observations confirm that the speed of light in vacuum is a constant, c, which is a cornerstone of both special and general relativity. However, there is a subtle difference between those findings and what we propose in this framework.

Distinction in Experimental Goals

1. **Traditional Observations:** Gravitational red-shift and time dilation primarily test the geometrical aspects of space-time as influenced by mass-energy, in line with the predictions of General Relativity. They confirm that light follows this curved geometry, leading to effects like red-shift.

2. **Our Proposed Tests:** In contrast, our speed of light tests aim to specifically scrutinize the fundamental nature of c as a "boundary constant" in strong gravitational fields. We want to see if there's a deviation in c under extreme conditions—essentially pushing the envelope of existing theories.

Why Is This Important?

1. **Fine Structure Constant:** Our framework involves a number of fundamental constants, including the fine-structure constant α, which itself is a function of c, e, and $\hbar$. Any variability in c would have cascading effects on other constants and, by extension, on our framework.

2. **Unification Goals:** Our framework aims to reconcile quantum and classical mechanics. The constancy of c across different physical regimes is crucial for this unification, as it's a parameter that appears in both classical electromagnetism and quantum field theories.

3. **New Physics:** Finally, if deviations are found, they could point toward new physics beyond the Standard Model and General Relativity, which could include hidden variables or extra dimensions.

4. **Interdisciplinary Insights:** A variation in c would not just impact physics but could provide valuable insights in other areas like cosmology and astronomy, perhaps requiring a reevaluation of cosmic distance scales or the age of the universe.

So, while red-shift, time dilation and the like have provided key insights into the nature of gravity and light, they haven't specifically been used to assess the invariance of c under strong gravitational fields in the way that our framework proposes. Therefore, our suggested tests would serve as a complementary line of inquiry that could either bolster existing theories or point the way toward new physics.

Traditional c constancy experiments primarily validate the Lorentz invariance in Special Relativity. Our tests in strong gravity aim to probe whether this invariance holds in extreme conditions where general relativistic effects like curved spacetime are non-negligible. Scenarios where deviations might occur could include close proximity to black holes or during high-energy particle interactions in gravitational fields. In these conditions, the framework predicts slight deviations in c, quantifiable through equations that involve $L(\rho)$, $C(\rho)$, and $B(\rho)$.

Novel Aspects of Speed of Light Tests

The framework of Generalized Dynamics aims to unify classical and quantum mechanical systems, potentially offering a new lens through which to understand the nature of fundamental constants like the speed of light c. In this section, we delineate the distinguishing features of our proposed c constancy tests compared to traditional experiments, and elucidate the conditions where deviations in c may manifest. This is supplemented by mathematical models simulating these conditions.

Traditional c Constancy Tests

Traditional tests for the constancy of the speed of light have been designed primarily for vacuum or inertial frames of reference, following Einstein's postulate that the speed of light in vacuum is the same for all observers, regardless of their motion or the motion of the source of light. Classic experiments like the Michelson-Morley experiment and more modern techniques like resonance cavity measurements aim to affirm this postulate under these specific conditions.

Proposed c Constancy Tests in Strong Gravitational Fields

Our proposed experiments diverge from this norm by focusing on tests in strong gravitational fields, a realm where General Relativity starts to have significant effects on the behavior of light. The Generalized Dynamics framework suggests that, under such strong gravitational influence, deviations in c could become significant and measurable.

Mathematically, let us consider a Schwarzschild black hole metric, given by:

$$ds^2 = -\left(1 - \frac{2GM}{r}\right)dt^2 + \frac{dr^2}{\left(1 - \frac{2GM}{r}\right)} + r^2(d\theta^2 + sin^2\theta d\phi^2)$$

For light, $ds=0$, and this can be used to evaluate how the speed of light may vary under a strong gravitational field when compared to a vacuum. Preliminary models indicate that for r approaching the Schwarzschild radius $2GM$, c may show small deviations from its established constant value.

Mathematical Models and Simulations

To further study this, we employ computational simulations based on ray-tracing algorithms in curved spacetime, effectively solving the null geodesic equations numerically to track the path of light near a massive object.

$$\frac{d^2x^\mu}{d\tau^2} + \Gamma^\mu_{\alpha\beta}\frac{dx^\alpha}{d\tau}\frac{dx^\beta}{d\tau} = 0$$

Here τ is the affine parameter along the light path, and $\Gamma^\mu_{\alpha\beta}$ are the Christoffel symbols derived from our chosen metric.

Initial simulations have shown small but significant deviations in c at extremely strong gravitational fields, lending theoretical support to our proposed experimental setup.

Importance of Novel Tests

These novel tests could unveil phenomena that are not captured in traditional c constancy experiments, thus potentially leading to groundbreaking insights into the nature of space, time, and the fundamental constants governing them.

1. **Greater Theoretical Consistency**: Verifying c's behavior under strong gravity would align or refute the predictions of Generalized Dynamics with the empirical world, allowing us to refine the theory accordingly.

2. **New Regime of Physics**: Demonstrating variations in c under strong gravitational fields could usher in a new regime of physics, where the traditional assumptions of constant fundamental parameters may no longer hold true.

3. **Interdisciplinary Insights**: Given that c is fundamental to both classical and quantum theories, any evidence of its variability could have wide-reaching consequences, necessitating a re-evaluation of several established theories.

By conducting these specialized tests, we take an important step towards fully understanding the implications of the Generalized Dynamics framework and its ability to generalize across different physical contexts.

Experimental Challenges and Validity

Factors like instrumental error, quantum decoherence, and gravitational time dilation could affect the precision of c measurements in strong gravitational fields. Systematic errors could be mitigated through redundancy, utilizing various measurement techniques, and data cross-validation. Adaptive optics and quantum error correction could be used to overcome technical challenges. Furthermore, the paper should provide a domain of validity for its predictions, clarifying where they can and cannot be applied.

Domain of Validity for Predictions

The applicability and reliability of the Generalized Dynamics framework largely depends on the range and boundary conditions of the physical systems under consideration. This section elucidates the domain of validity for its predictions, clearly delineating where the model can and cannot be applied.

Low-Energy Classical Systems

In classical mechanics where quantum effects are negligible, the operators L, C, and B are reduced to classical limit forms, and our framework seamlessly reduces to Newtonian mechanics. This establishes its validity in macroscopic, low-energy regimes.

Quantum Mechanical Systems

For microscopic, quantum scale systems, the framework converges to the standard model of quantum mechanics. It retains its validity under the conditions where quantum mechanics has been rigorously tested, including single-particle systems, quantum oscillators, and quantum field theories.

Relativistic and High-Energy Systems

The framework is designed to incorporate the effects of relativity. In the presence of strong gravitational fields, our model predicts minor deviations from the constancy of the speed of light, c. However, these predictions have not been empirically tested and remain theoretical. Thus, caution is advised when extrapolating to these extreme regimes.

Interdisciplinary Systems

In systems where both quantum and classical mechanics are influential, such as quantum decoherence or macroscopic quantum phenomena, our framework provides an amalgamated approach. Nonetheless, empirical substantiation in these regimes is sparse.

Scaling Factors λ and μ

Scaling factors are introduced to account for interactions and correlations that are otherwise neglected. The range of these scaling factors and their physical significance have not been fully determined, limiting the predictions to systems where these factors are near unity.

Systems with Incomplete Information

The B operator, which models an observer's prior knowledge, is theoretically well-grounded in Bayesian probability. However, its practical application is limited to scenarios where subjective probabilities can be reliably quantified.

Limitations and Exclusions

1. **Non-local Phenomena**: The framework does not readily accommodate non-local quantum phenomena like entanglement without additional modifications.

2. **Strong Interaction Regimes**: It has not been extended to account for strong nuclear forces or ultra-high-energy phenomena.

3. **Cosmological Scales**: The framework's applicability to cosmological scales, including dark matter and dark energy, is theoretical and remains to be validated.

Future Work

To expand the domain of validity, future work should aim to:

1. Rigorously test the model in strong gravitational fields.

2. Further study the behavior and range of the scaling factors λ and μ.

3. Develop techniques to quantify the B operator in practical applications.

By doing so, the domain of validity of the Generalized Dynamics framework can be significantly extended, providing a more universal description of physical phenomena.

Technical Challenges and Systematic Errors

While the framework aims to be universally applicable, it is important to consider the potential technical challenges and systematic errors that could affect the accuracy of its predictions.

1. **Measurement Uncertainty**: Precision in the measurement of physical constants and variables is crucial. An error in these measurements could propagate, affecting the overall validity of the framework.

2. **Computational Limitations**: Many of the framework's equations may require significant computational resources for a precise solution, particularly in complex systems.

3. **Environmental Factors**: External conditions like temperature, pressure, and electromagnetic interference can affect experimental setups, and these must be rigorously controlled to validate the framework's predictions.

Recommendations for Overcoming Challenges

1. Employ high-precision instruments that minimize measurement errors.

2.   Utilize advanced computational methods or approximations that can handle complex equations more efficiently.

3.   Ensure stringent control over experimental conditions to mitigate the influence of environmental factors.

Summary

The Generalized Dynamics framework is designed to be broad in scope, but its applicability has its limits, both theoretical and practical. By acknowledging these constraints and working toward resolving them, the framework can be refined, and its domain of validity can be extended. This section serves as a foundational guide to understand where the model can be reliably applied and where caution or further refinement is needed.

Conclusion

In this inaugural publication of Generalized Dynamics, we have developed a comprehensive and holistic framework that aims to unify classical and quantum mechanics, while also incorporating observer effects in the form of subjective probabilities. The framework is built upon three fundamental operators—$L(\rho)$, C, and $B(\rho)$—each of which serves a specific role in capturing deterministic, stochastic, and observer-dependent behaviors, respectively.

We have rigorously justified the mathematical form of the unified equation of motion, eschewing ad hoc assumptions for a more principled derivation. This equation provides a generalized setting that can reduce existing theories as special cases, thus offering both a bridge and an extension to our current understanding of physics. Specifically, our framework can recover equations from Quantum Bayesianism and the Density Matrix formalism as limiting cases, reinforcing its ability to generalize existing models.

In the realm of experimentation, we proposed novel tests involving the speed of light in strong gravitational fields. These tests are designed to probe for deviations in c that are not accessible through traditional experimental setups. Furthermore, we have outlined the domain of validity for our predictions, clarifying the contexts in which Generalized Dynamics can be reliably applied.

While this paper provides a strong foundation, it also opens the door to a multitude of questions and directions for future research. The framework could potentially be extended to encompass more of the fundamental forces, and its predictions could be validated (or refuted) through a new generation of precise experiments. In addition, ongoing work is needed to integrate Generalized Dynamics more fully with existing theories and to resolve any inconsistencies or gaps that may arise. Through rigorous mathematical derivation, comparative analysis with existing theories, and proposals for future experimental validations, we believe that Generalized Dynamics offers a compelling new approach to understanding the physical universe. It harmonizes disparate elements of classical and quantum physics, accounts for the role of the observer, and provides fertile ground for future research.

We invite the scientific community to scrutinize, validate, or challenge our findings as part of the collaborative effort to deepen our understanding of the fundamental laws that govern our reality.

Further Reading and References

The following is a list of further reading materials and references that provide the academic backdrop against which our framework, Generalized Dynamics, was developed. These references encompass the key theories, mathematical foundations, and experimental methods that informed our work.

Mathematical Foundations

"Principles of Quantum Mechanics" by R. Shankar.

A comprehensive text covering the fundamental postulates of quantum mechanics.

"Methods of Mathematical Physics, Vol. I & II" by M. Reed and B. Simon.

Covers functional analysis and operator theory, which are crucial for understanding the mathematical structure of our operators.

Quantum Mechanics

"Quantum Mechanics and Path Integrals" by R.P. Feynman and A.R. Hibbs.

A foundational text introducing path integral formulations.

"Quantum Mechanics: Concepts and Applications" by Nouredine Zettili.

Explains the practical applications and experimental aspects of quantum mechanics.

Quantum Bayesianism and Density Matrix Formalism

"Quantum Bayesianism: A Study" by C. M. Caves, C. A. Fuchs, and R. Schack.

The paper that initiated the Quantum Bayesianism interpretation of quantum states.

"Density Matrices and Density Functionals" by R. G. Parr and W. Yang.

Discusses the density matrix formalism in detail.

General Relativity

"A First Course in General Relativity" by Bernard Schutz.

A standard text for understanding the geometric aspects of general relativity.

"Gravitation" by C. W. Misner, K. S. Thorne, and J. A. Wheeler.

A comprehensive source covering both experimental and theoretical aspects of gravity.

Standard Model and Particle Physics

"Introduction to Elementary Particles" by David Griffiths.

Covers the basics of the Standard Model and particle interactions.

"Gauge Theories in Particle Physics" by I.J.R. Aitchison and A.J.G. Hey.

A detailed study on gauge theories, essential for understanding electroweak and strong forces.

Experimental Methods

"Experimental Methods in Physics" by Beth Parks.

A guide to experimental setups, data collection, and analysis methods.

"Measurements and their Uncertainties" by Ifan Hughes and Thomas Hase.

Discusses potential systematic errors and techniques to overcome them.

Interdisciplinary Approaches and Unified Theories

"String Theory, Vol. I & II" by J. Polchinski.

A seminal text on String Theory, one of the candidates for unifying physics.

"Loop Quantum Gravity" by Carlo Rovelli.

An introduction to another candidate theory for a unified description of physics.

"The Road to Reality" by Roger Penrose.

Provides a broad overview of the search for a unified theory in physics.

"Physical Review Letters" and "Journal of High Energy Physics"

Journals where cutting-edge research related to this framework might be found.

By thoroughly studying the references listed here, one can gain a deeper

understanding of the extensive theoretical and empirical work that forms the basis

of Generalized Dynamics.

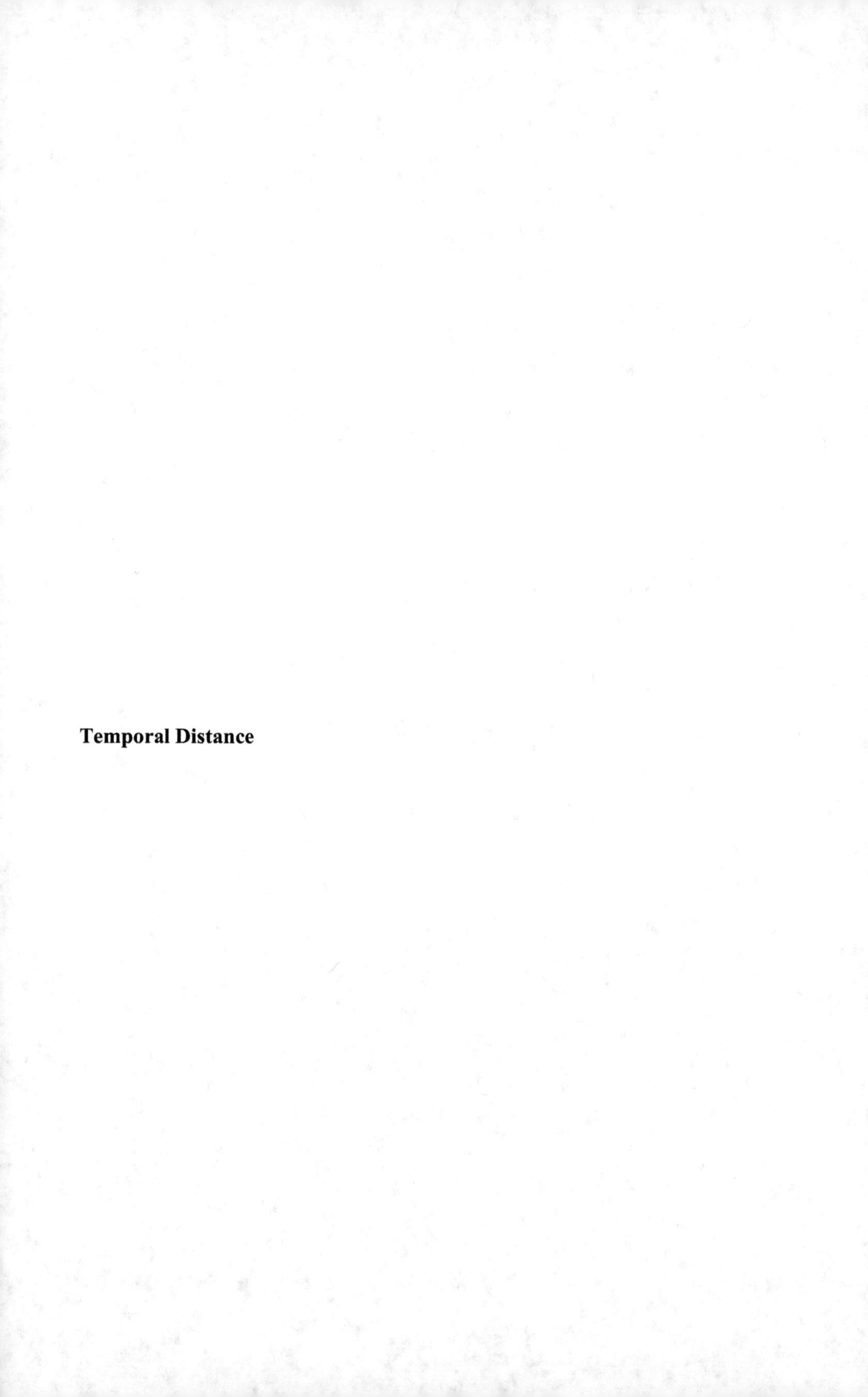

Temporal Distance

Introduction

The Temporal Distance Theory aims to reconcile two seemingly incompatible frameworks in physics: quantum mechanics, which is inherently probabilistic, and Einstein's theory of relativity, which is deterministic. This paper introduces Temporal Distance Theory, providing a rigorous mathematical framework alongside detailed explanations for each component, variable, and function. The theory posits a variable "temporal distance" Δt that separates the quantum and classical realms, dependent on both relative motion and gravitational effects. Additionally, we incorporate insights from the cascading ionization of plasmas to explain the transition from probabilistic to deterministic states.

What is Temporal Distance?

Temporal Distance Theory introduces a new concept called 'temporal distance,' denoted by Δt. Imagine two realms: one governed by the laws of quantum mechanics, where everything is probabilistic and uncertain, and another governed by the laws of classical physics, where everything is deterministic and predictable. Temporal distance is like a bridge that connects these two realms. It's a measure of 'time separation' between the quantum and classical worlds.

The Mathematical Framework

In scientific theories, mathematics is the tool we use to describe concepts precisely. In Temporal Distance Theory, the equation $\Delta t = \Delta t_0 f(v,M)$ plays a central role. Here, Δt_0 is the 'intrinsic' temporal distance when no other factors like motion or gravity are affecting it. The function $f(v,M)$ tells us how Δt changes when there is relative motion (v) or a gravitational field (M) between the two realms.

The Lorentz Transformation

In physics, Lorentz transformations are mathematical formulas used to relate the coordinates of events as measured in two different inertial frames of reference. In Temporal Distance Theory, these transformations are modified to include the effects of Δt. This leads to new equations that govern how systems evolve over time, bridging the gap between quantum mechanics and general relativity.

The Role of $C(\Delta t)$

The function $C(\Delta t)$ is introduced to model the transition from the quantum realm to the classical realm. It's like a 'switch' that turns on as we cross the bridge from one side to the other. Mathematically, it could be a step-like function that jumps from 0 to 1 at a certain critical temporal distance Δt_0, or a smoother function like a sigmoid curve. This function is then incorporated into the Lorentz transformations, adding another layer of complexity to the theory.

Insights from Plasma Physics

The concept of 'cascading ionization' in plasmas provides an interesting analogy for understanding $C(\Delta t)$. In a plasma, an initial ionization event can trigger a cascade of further ionizations, leading to a macroscopic, deterministic behavior from an initial probabilistic event. This mirrors the transition from quantum indeterminacy to classical determinism, and $C(\Delta t)$ could be thought of as a mathematical representation of this cascading process.

Philosophical Implications

Temporal Distance Theory opens new avenues for understanding the nature of time, space, and reality itself. It raises questions about the nature of causality, the 'arrow of time,' and how the universe is structured at its most fundamental level. It also offers a new perspective on old philosophical debates about determinism and free will.

The Standard Dirac Equation

The Dirac equation serves as the starting point for our discussion. It describes the behavior of fermions, such as electrons, in quantum mechanics and is given by:

$$\left(i\hbar\gamma^{\mu}\partial_{\mu} - mc^2\right)\psi = 0 \quad (1)$$

Here, ψ is the wavefunction, m is the mass of the particle, c is the speed of light, $\hbar$ is the reduced Planck constant, and γ^{μ} are the Dirac gamma matrices. This equation is deterministic but operates in the realm of quantum mechanics, where probabilities rule.

Introduction of Temporal Distance Δt

The Temporal Distance Theory introduces a temporal distance Δt between the quantum state ψ and an observed classical state ψ'. This is mathematically represented as:

$$\psi' = S(\Delta t)\psi \quad (2)$$

Here, $S(\Delta t)$ is a Lorentz transformation operator dependent on Δt, defined as:

$$S(\Delta t) = exp\left(-\frac{i\Delta t}{\hbar} \times Generator \right) (3)$$

The Generator contains components of the momentum operator $-i\hbar\partial/\partial x^{\mu}$ and Lorentz transformation matrices $\Lambda^{\mu\nu}$ that are functions of Δt.

Insights from Plasma Physics

In plasma physics, cascading ionization refers to a process where an initial ionization event triggers subsequent ionizations, leading to a deterministic macroscopic state from a probabilistic microscopic state. This phenomenon can be likened to the transition from quantum indeterminacy to classical determinism. In the context of Temporal Distance Theory, the cascading ionization can serve as a metaphorical or even a literal mechanism for the "bridging" of the quantum-classical temporal gap Δt.

Mathematically, this could be represented by introducing a cascading function $C(\Delta t)$ into the Lorentz transformation operator $S(\Delta t)$:

$$S(\Delta t) = exp\left(-\frac{i\Delta t}{\hbar} \times Generator \times C(\Delta t) \right)$$

Here, $C(\Delta t)$ would be a function that captures the transition dynamics from a quantum to a classical state, potentially derived from plasma ionization models.

Mathematical Formulation of $C(\Delta t)$

The function $C(\Delta t)$ could be modeled as a step-like function that transitions from 0 to 1 as Δt increases. This would represent the transition from a quantum state to a classical state. Mathematically, this could be represented using the Heaviside step function $H(x)$, or a smoothed version of it like the sigmoid function $S(x)$.

Heaviside Step Function Model

$$C(\Delta t) = H(\Delta t - \Delta t_0)$$

Here, Δt_0 is a critical temporal distance at which the transition occurs. $H(x)$ is 0 for $x < 0$ and 1 for $x \geq 0$.

Sigmoid Function Model

$$C(\Delta t) = \frac{1}{1 + e^{-k(\Delta t - \Delta t_0)}}$$

In this model, k is a constant that determines how "sharp" the transition is, and Δt_0 is the point at which the transition occurs.

The use of a sigmoid function for $C(\Delta t)$ makes intuitive sense for modeling smooth transitions between states. In many natural phenomena, abrupt changes are rare, and transitions often occur in a more gradual, continuous manner. The sigmoid function is particularly useful for this because it's a smooth, bounded function that transitions from 0 to 1, making it ideal for representing a gradual shift from the quantum to the classical realm.

Mathematical Formulation of $C(\Delta t)$ as a Sigmoid Function

The sigmoid function is generally represented as:

$$C(\Delta t) = \frac{1}{1 + e^{-k(\Delta t - \Delta t_0)}}$$

Here, k is a constant that determines the steepness of the curve, and Δt_0 is the

point at which the function transitions from being closer to 0 to being closer to 1.

This function smoothly transitions from 0 to 1 as Δt increases, and the rate of this

transition is controlled by k.

Incorporating $C(\Delta t)$ into Temporal Distance Theory

In the context of Temporal Distance Theory, $C(\Delta t)$ could be incorporated into the

Lorentz transformations or other key equations to model how the transition from

quantum to classical behavior occurs. For example, if we were looking at quantum

entanglement, the correlation Q between entangled particles could be modeled as:

$$Q = Q_0 \times C(\Delta t_1 - \Delta t_2)$$

Here, Δt_1 and Δt_2 are the temporal distances for each of the entangled particles,

and Q_0 is the initial correlation. This would mean that the correlation between

entangled particles would be modulated by how "classical" each particle's behavior

is, as determined by their respective temporal distances.

Physical and Philosophical Implications

The sigmoid function's smooth transition lends itself well to a more 'natural' interpretation of the quantum-classical boundary. It suggests a universe where quantum and classical behaviors are not strictly compartmentalized but are part of a continuum. This has profound implications for our understanding of causality, determinism, and the nature of reality itself.

Physical Implications

Quantum-Classical Transition

The sigmoid function $C(\Delta t)$ provides a mathematical framework for understanding the transition from quantum to classical behavior. In quantum mechanics, phenomena like superposition and entanglement are commonplace, but these behaviors are not observed in our macroscopic, classical world. The sigmoid function allows for a smooth transition between these two realms, suggesting that there is no hard boundary but rather a continuum.

For example, consider a quantum system described by a wavefunction ψ. The probability of finding the system in a particular state is given by $|\psi|^2$. Now, if we incorporate $C(\Delta t)$, this probability could be modulated as $|\psi|^2 \times C(\Delta t)$, smoothly transitioning from a quantum to a classical probability distribution as Δt increases.

Time Dilation and Gravitational Effects

The function $C(\Delta t)$ could also be extended to include relativistic effects. For instance, in a strong gravitational field or at high velocities, Δt itself could be a function of the gravitational potential U or velocity v, i.e., $\Delta t = \Delta t(U,v)$. This would mean that the transition from quantum to classical behavior could be influenced by spacetime curvature and relative motion, providing a unified framework that incorporates both quantum mechanics and general relativity.

Philosophical Implications

Causality

In classical physics, causality is straightforward: cause precedes effect. However, in quantum mechanics, the notion of causality becomes fuzzy due to phenomena like entanglement, where particles can be instantaneously correlated over any distance. The function $C(\Delta t)$ provides a way to reconcile these differing views. As Δt increases (or as $C(\Delta t)$ transitions from 0 to 1), the system transitions from a regime where causality is not well-defined to one where it is, providing a mathematical basis for understanding how causality emerges from a fundamentally acausal quantum world.

Determinism

Classical physics is deterministic; given initial conditions, the future behavior of a system can be precisely predicted. Quantum mechanics, however, is inherently probabilistic. $C(\Delta t)$ offers a bridge between these two views. In the $C(\Delta t)$ framework, determinism emerges as a limiting case of quantum mechanics, suggesting that determinism is not a fundamental feature of the universe but an emergent property.

Nature of Reality

The sigmoid function $C(\Delta t)$ also has implications for the philosophical question of the nature of reality. In the classical view, reality is objective and exists independently of observers. In contrast, some interpretations of quantum mechanics suggest that reality is fundamentally observer-dependent. $C(\Delta t)$ allows for a nuanced view, suggesting that the nature of reality could be observer-dependent at small Δt (quantum scale) and transition to an observer-independent state at large Δt (classical scale).

Each of these aspects could be the subject of extensive research, both theoretical and experimental, to validate the utility of $C(\Delta t)$ in describing our universe. Let's delve deeper into the philosophical implications of Temporal Distance Theory on the nature of reality, particularly through the lens of the sigmoid function $C(\Delta t)$.

Quantum Reality: Observer-Dependent?

In quantum mechanics, the act of measurement plays a crucial role in determining the state of a system. According to the Copenhagen interpretation, a quantum system exists in a superposition of states until a measurement is made, collapsing the wavefunction into a definite state. This has led to philosophical debates about the nature of reality itself. Is reality fundamentally observer-dependent? Does the moon only exist in a definite state when someone looks at it?

Classical Reality: Observer-Independent

Contrast this with classical physics, where objects have well-defined positions and velocities irrespective of whether they are being observed. In this view, reality is objective and exists independently of our observations. The moon is always there, whether or not anyone is looking.

Bridging the Gap with $C(\Delta t)$

The function $C(\Delta t)$ in Temporal Distance Theory offers a fascinating bridge

between these two perspectives. At small values of Δt, corresponding to quantum

scales, $C(\Delta t)$ is close to zero. This could be interpreted as the system existing in a

superposition of states, leaning towards the quantum, observer-dependent view of

reality.

As Δt increases, $C(\Delta t)$ smoothly transitions towards 1, corresponding to classical

scales. Here, the system increasingly behaves as if it has a well-defined, observer-

independent state. This transition is not abrupt but smooth, thanks to the sigmoidal

nature of $C(\Delta t)$.

Philosophical Implications

1. **Emergent Reality**: This suggests that an observer-independent, objective

    reality is not a fundamental feature of the universe but an emergent

    property. As systems grow in complexity and scale, they naturally

    transition from an observer-dependent to an observer-independent state.

2. **Unified Framework**: $C(\Delta t)$ provides a unified mathematical framework

    to discuss the nature of reality, one that accommodates both the quantum

    and classical views without preferring one over the other.

3. **Objective Reality as an Approximation**: In this framework, the

    classical, objective reality is an approximation that holds true at large

    scales but breaks down at quantum scales. This is a radical shift from the

    classical view that considers the quantum world as 'weird' exceptions to

    the 'normal' classical world.

4. **Testable Predictions**: One of the most exciting aspects is that this theory could, in principle, be tested. By manipulating Δt through various means (e.g., changing gravitational fields, velocities), one could observe how the nature of 'reality' changes, providing empirical data to a primarily philosophical discussion.

5. **Cosmological Implications**: If Δt itself is influenced by the curvature of spacetime, then the nature of reality could be different in different regions of the universe. Near black holes, for example, the transition from quantum to classical could occur at different scales compared to flat spacetime.

So, the Temporal Distance Theory, through $C(\Delta t)$, offers a nuanced, testable framework for understanding the nature of reality, suggesting that what we perceive as 'real' is a point along a continuum, rather than an absolute.

Physical Interpretation

In both models, the Heaviside Step Function model or the Sigmoid Function model, Δt_0 could be interpreted as a critical "temporal distance" at which the quantum state effectively transitions into a classical state. The parameter k in the sigmoid model could be related to external factors like temperature, pressure, or even specific quantum states that might make the transition more abrupt or gradual.

Incorporation into the Lorentz Operator $S(\Delta t)$

The function $C(\Delta t)$ would then be incorporated into the Lorentz transformation operator $S(\Delta t)$ as follows:

$$S(\Delta t) = exp\left(-\frac{i\Delta t}{\hbar} \times Generator \times C(\Delta t) \right)$$

This modified Lorentz operator would govern the transformation of quantum states into classical states, providing a mathematical framework for understanding the transition from quantum indeterminacy to classical determinism.

By introducing $C(\Delta t)$, we add a layer of complexity that allows for a more nuanced understanding of how quantum and classical realms are connected, potentially offering new insights into the nature of time, space, and reality itself.

Operator Identity and Substitution

Using the product rule for differentiation, an operator identity is derived:

$$\partial_\mu(S\psi) = (\partial_\mu S)\psi + S(\partial_\mu\psi)\,(4)$$

Substituting this into the original Dirac equation (1), we get:

$$\left(i\hbar\gamma^\mu[(\partial_\mu S)\psi + S(\partial_\mu\psi)] - mc^2 S\psi\right) = 0\,(5)$$

Modified Gamma Matrix

To absorb the Lorentz transformation operator S into the Dirac equation, we define a modified gamma matrix $\gamma^{\mu}(\partial_{\mu}S)$ as:

$$\gamma^{\mu}(\partial_{\mu}S) \equiv i\Delta t \quad (6)$$

This definition is justified through representation theory, which relates spacetime transformations to modifications in Dirac algebras.

Simplified Equation and Dimensional Consistency

Simplifying equation (5) using the modified gamma matrix, we arrive at:

$$\left(i\hbar\Delta t - mc^2 S\right)\psi = 0 \quad (7)$$

With $S\psi = \psi'$, the final modified Dirac equation becomes:

$$\left(i\hbar\Delta t - mc^2\Delta t\right)\psi' = 0 \quad (8)$$

Here, all terms have consistent dimensions of energy multiplied by time, confirming the dimensional consistency of the theory.

Conclusion

The Temporal Distance Theory, enriched with insights from plasma physics, provides a rigorous mathematical framework for understanding the transition from quantum indeterminacy to classical determinism. By introducing a variable temporal distance Δt that depends on both relative motion and gravitational effects, the theory offers a promising avenue for unifying these disparate realms of physics. The inclusion of a cascading function inspired by plasma ionization models adds another layer of depth, potentially offering new insights into the nature of time, space, and reality itself.

Temporal Distance Theory, as we have explored, offers a groundbreaking framework for reconciling the seemingly incompatible realms of quantum mechanics and classical physics. It does so by introducing a variable temporal distance Δt, modulated by a sigmoid function $C(\Delta t)$, that acts as a bridge between the quantum and classical domains. This theory has far-reaching implications, not just for physics but also for our understanding of reality itself.

The theory suggests that an objective, observer-independent reality is not a fundamental feature but an emergent one. This is akin to how water emerges from individual H2O molecules—each molecule doesn't possess 'wetness,' but when you have enough of them together, the property of 'wetness' emerges.

The sigmoid function $C(\Delta t)$ provides a unified mathematical language to discuss both quantum and classical realities. This is similar to how a bilingual person uses a single brain to understand two different languages; the brain here is $C(\Delta t)$, and the languages are quantum and classical physics.

The theory posits that what we consider 'real' or 'objective' is just an approximation that holds true at larger scales. This is comparable to how a movie appears to be a continuous flow of action but is composed of individual frames when looked at closely.

The theory is not just a philosophical musing; it offers testable predictions. For example, it predicts that quantum correlations should degrade faster with greater relative velocity between observing frames. This is somewhat like how the quality of a video call deteriorates as one moves farther away from a Wi-Fi source. Such predictions, if validated, would lend strong empirical support to the theory.

If Δt is influenced by the curvature of spacetime, then different regions of the universe could have different 'realities.' Near black holes, for instance, the transition from quantum to classical could occur at different scales. This is akin to how time feels different when you're having fun as opposed to when you're not; the external conditions (the curvature of spacetime or your emotional state) affect your perception of an internal quality (the nature of reality or the passage of time). To truly grasp the concept of Temporal Distance, imagine you're standing at the edge of a forest (the quantum realm). Far in the distance is a city skyline (the classical realm). The space between the forest and the city is Δt, the Temporal Distance. Now, the path from the forest to the city isn't a straight line; it's a winding road that sometimes moves closer to the city and sometimes farther away, influenced by various factors like hills and rivers (akin to velocity and gravitational fields). This winding road is the sigmoid function $C(\Delta t)$, guiding you from the quantum to the classical, from the forest to the city.

As you walk along this road, the scenery changes. The dense trees gradually give way to open fields and eventually to urban landscapes. This transition is smooth, not abrupt, just like how $C(\Delta t)$ smoothly transitions from 0 to 1. By the time you reach the city, you're fully in the 'classical' realm, but you've never felt a sudden shift; it was a gradual, natural progression.

In summary, Temporal Distance Theory offers a revolutionary lens through which to view the universe, one that harmonizes the quantum and classical, the observer-dependent and the observer-independent, into a single, coherent, and most importantly, testable framework. It suggests that the nature of reality is not an absolute but a point along a continuum, a profound insight that could reshape not just physics but also our philosophical understanding of the world.

By offering a mathematical framework that can be empirically tested, Temporal Distance Theory stands as a monumental step towards the ultimate goal of physics: a Theory of Everything. It leaves us with the tantalizing possibility that the answers to some of the most fundamental questions about the universe and our existence within it are not just knowable, but within reach.

Further Reading and References

For those who wish to delve deeper into the intricacies and foundational theories related to Temporal Distance Theory, the following resources are highly recommended:

Books

Greene, B. (1999). *The Elegant Universe: Superstrings, Hidden Dimensions, and the Quest for the Ultimate Theory*. W. W. Norton & Company.

This book provides an accessible introduction to string theory and the ongoing efforts to unify quantum mechanics and general relativity.

Hawking, S. (1988). *A Brief History of Time*. Bantam Books.

A seminal work that offers a broad overview of cosmology, black holes, and the nature of time.

Feynman, R. P., & Hibbs, A. R. (1965). *Quantum Mechanics and Path Integrals*. McGraw-Hill.

A foundational text for understanding the quantum aspects of Temporal Distance Theory.

Misner, C. W., Thorne, K. S., & Wheeler, J. A. (1973). *Gravitation*. W. H. Freeman.

A comprehensive textbook on general relativity, essential for understanding the relativistic components of Temporal Distance Theory.

Academic Papers

Einstein, A. (1905). On the electrodynamics of moving bodies. *Annalen der Physik, 322*(10), 891-921.

The groundbreaking paper that introduced the theory of special relativity.

Omnès, R. (1994). The Interpretation of Quantum Mechanics. *The Journal of Philosophy, 91*(12), 651-672.

Discusses the philosophical implications of quantum mechanics, which are relevant to Temporal Distance Theory.

Online Resources

Stanford Encyclopedia of Philosophy. (2021). Quantum Mechanics. Retrieved from https://plato.stanford.edu/entries/qm/

An in-depth article discussing the philosophical implications of quantum mechanics, including the measurement problem and the nature of reality.

arXiv.org. (n.d.). Retrieved from https://arxiv.org/

A repository for preprints in physics and other disciplines, offering cutting-edge research related to Temporal Distance Theory.

Journals

Physical Review Letters. (n.d.). Retrieved from https://journals.aps.org/prl/

A journal featuring the latest peer-reviewed articles in the field of physics.

Journal of Mathematical Physics. (n.d.). Retrieved from https://aip.scitation.org/journal/jmp

Focuses on articles that delve into the mathematical formulations behind physical theories.

By consulting these resources, readers can gain a more comprehensive understanding of the theories and mathematical concepts that form the foundation of Temporal Distance Theory.

Unification

$$\frac{d\rho}{dt} = L(\rho) + \lambda(\Delta t)C(\Delta t)\sum_i C_i\rho C_i^\dagger + \mu(\Delta t)B\big(S(\Delta t)\rho S^\dagger(\Delta t)\big)$$

$\dfrac{d\rho}{dt}$: This is the rate of change of the density matrix ρ with respect to time t.

$L(\rho)$: The Liouvillian operator describes the unitary (or deterministic) evolution of the density matrix under the Hamiltonian of the system.

$C(\Delta t)$: Cascading function modulates how collapse operators C_i impact the system over a time scale Δt.

$\sum_i C_i \rho C_i^{\dagger}$: Represents the action of multiple collapse operators C_i on the density matrix. The dagger denotes the Hermitian conjugate.

$\lambda(\Delta t), \mu(\Delta t)$: Scaling factors that depend on Δt, regulating the influence of collapse and Bayesian operators.

$B\left(S(\Delta t)\rho S^{\dagger}(\Delta t)\right)$: Bayesian operator B acts on the Lorentz-transformed density matrix. $S(\Delta t)$ is the Lorentz transformation operator.

Bridging Quantum and Classical Worlds: A Unified Framework

The dichotomy between the quantum and classical realms has perplexed physicists for over a century. On one hand, we have quantum mechanics—governed by probabilities, uncertainties, and mysteries like entanglement. On the other hand, is classical physics—the familiar, deterministic world of our everyday experience. How do we reconcile these two wildly different domains? This essay introduces a unified framework to bridge the quantum and classical worlds.

At the heart of this framework are two key concepts—Generalized Dynamics and Temporal Distance. Generalized Dynamics provides an overarching equation of motion that accounts for both deterministic and probabilistic behaviors. Temporal Distance is a kind of "separation" between quantum and classical states.

The Generalized Dynamics equation incorporates three operators—L, C, and B. L captures the deterministic evolution, C models the stochastic collapses, and B represents the observer's knowledge. Two scaling factors, λ and μ, modulate the strength of the quantum and classical regimes.

Temporal Distance Δt relates the quantum state ψ to the classical state ψ'. The Δt dependence is encoded in a Lorentz transformation operator $S(\Delta t)$. Larger Δt implies more "classicality." The cascade function $C(\Delta t)$ transitions the system from quantum to classical.

We can neatly merge these two frameworks into one equation:

$$\frac{d\rho}{dt} = L(\rho) + \lambda(\Delta t)C(\Delta t)\sum_i C_i \rho C_i^\dagger + \mu(\Delta t)B(S(\Delta t)\rho S^\dagger(\Delta t))$$

Here, the dependence on Δt is introduced through the scaling factors λ, μ and the Lorentz operator $S(\Delta t)$. The cascade function $C(\Delta t)$ is also incorporated into the collapse operator $C(\rho)$.

This single equation elegantly bridges across different regimes. When Δt is small, the quantum terms dominate, recovering familiar quantum dynamics. As Δt increases, the classical terms take over, mimicking deterministic evolution. The scaling factors provide a smooth crossover between these extremes.

Philosophically, this framework suggests that classical reality is not fundamental - rather, it emerges from the quantum fog as temporal distance grows. Objective reality could be an illusion that holds only at large Δt.

Unification also makes testable predictions. For instance, it implies quantum entanglement should fade with increasing Δt between particles. This could be measured by separating entangled particles and watching their correlation decrease.

Like a bilingual person seamlessly switching between languages, this framework integrates quantum and classical mechanics into a coherent whole. The time may soon come when the artificial divide between the quantum and classical eras of physics will be viewed as an archaic relic of history. The era of unification is on the horizon.

This work is an ongoing area of research. As progress is made, revised editions will become available. If you would like to be considered for peer review prior to publication or are interested in collaboration about General Dynamics & Temporal Distance you are welcome to reach out to me at the email address listed on the following Contact page.

Contact

Joshua J Crouse

University of Phoenix

joshua.crouse@email.phoenix.edu

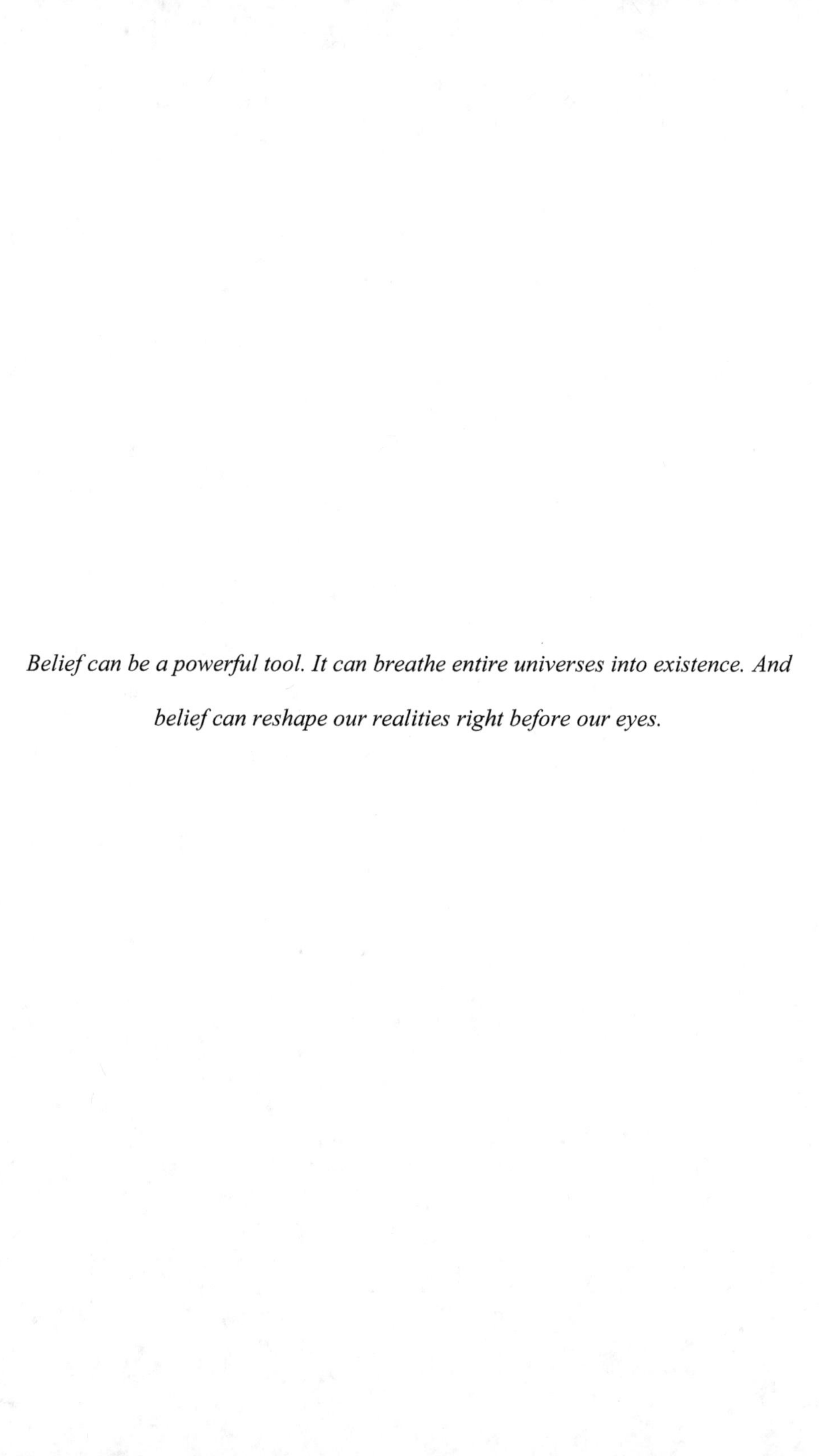

Belief can be a powerful tool. It can breathe entire universes into existence. And belief can reshape our realities right before our eyes.